DAMS
Water Tamers of the World

by BERNICE KOHN HUNT
illustrated by CHARLES ROBINSON

A Finding-Out Book  *Parents' Magazine Press • New York*

For Elizabeth and Emily

Library of Congress Cataloging in Publication Data
Hunt, Bernice Kohn.
  Dams: water tamers of the world.

  (A Finding-out book)
  Includes index.
  SUMMARY: Discusses the purpose, kinds, and planning of dams and includes chapters on "Dams through the Ages" and "When Dams Fail."
  1. Dams—Juvenile literature. [1. Dams]   I. Robinson, Charles, 1931-   II. Title.
TC540.H86   627'.8   76-26465
ISBNN 0-8193-0895-1

Text copyright © 1977 by Bernice Kohn Hunt
Illustrations copyright © 1977 by Charles Robinson
All rights reserved
Printed in the United States of America

CONTENTS

One	HOLDING BACK THE WATER	5
Two	DAMS THROUGH THE AGES	13
Three	ALL KINDS OF DAMS	20
Four	ELECTRICITY FROM WATER	28
Five	PLANNING A DAM	38
Six	A DAM IS BUILT: THE STORY OF CROTON DAM	47
Seven	WHEN DAMS FAIL	56
	INDEX	63

One
HOLDING BACK THE WATER

A long, long time ago, even before there were people on earth, the first dams stretched across rivers and streams. They were made of branches, mud, and stones; they had been built by beavers. These animals cut down trees by gnawing through the trunks with their teeth. Then they gnawed off branches, swam them out to the dam site, and fastened them in place with mud and rocks.

Today, beavers still build dams in exactly the same way—and for the same reason: to hold back water. Beavers dam up a stream until it is very deep. Then they build their homes, or lodges, right in the middle of the pond. A beaver lodge is like a castle with a deep moat all around it. When a beaver family is at home, it is safe from enemies.

Beaver dams may be very large. The largest one we know of is in Montana. It is more than 2,000 feet

long—equal to several very long city blocks. It is 12 feet high—as high as a one-story house—and 15 to 20 feet wide.

Beaver dams are very strong. They last for a long time because the animals keep them in good repair. They always need fresh branches for this work, and they like to feed on tender tree bark as well. After a while, they use up all of the trees near the shore. As soon as this happens, the beavers get busier than

ever. They quickly make their dams larger and flood some of the dry land. They work until the flood comes right up to the next row of trees. This is important for their safety, because beavers do not move quickly on land. As long as they can reach trees by swimming instead of walking, they can get new supplies without danger.

Like beavers, people build dams, too, and for the same main purpose—to hold back water. Their reasons for holding it back are different, though. The first dams were built to store water until it was needed. Once upon a time people had to live near streams so that they could have enough water. If the streams dried up, they were in serious trouble. After they learned to build dams, they could store water by damming up a stream when it was full. When the dry season came, they used the water they had saved.

A steady water supply made it possible for people to raise plants and to keep farm animals. For the first time, they were able to settle in one place instead of having to roam in search of water and food.

In more modern times, people built dams to hold back water to prevent floods. They also used them to make waterways deep enough for big boats.

Today, some dam-made lakes are used just for fun—for boating, swimming, and fishing. Above all, dams are used to make power—but we will learn more about that later on.

Dams often have more than one purpose. The main job of any dam is to hold back water, but a single dam can be useful in many ways. A dam that is good for more than one thing is called a *multi-purpose* dam. Most of the newest dams are multi-purpose. They are often made of concrete and are built by thousands of people. The oldest dams are made of sticks, and mud, and stones and are built by beavers. But because they are used both for moat-building and for flooding land, they are multi-purpose dams too.

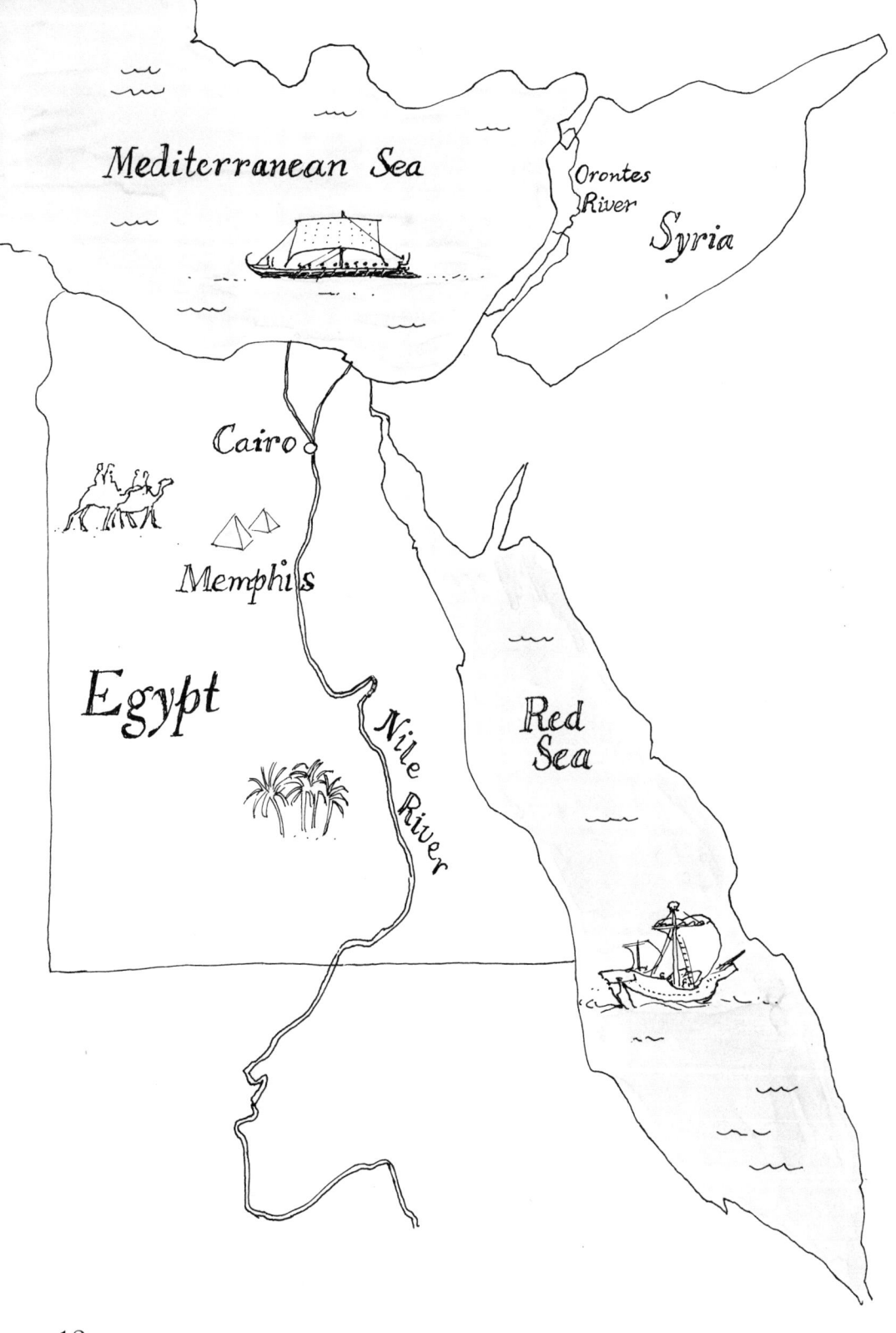

Two
DAMS THROUGH THE AGES

No one knows who the first human dam builders were, but we do have records of some very ancient dams. The earliest one was built on the Nile River in Egypt about 5,000 years ago, not long after the Stone Age. It supplied water to the capital of King Menes at Memphis.

Another dam, nearly as old, was built at Sadd-el-Kafara, near Cairo. Shortly after it was finished, a great flood of water turned it over and destroyed it. Its remains were discovered in 1885 by a German explorer, Georg Schweinfurth. He figured out that the dam had been about 350 feet long and had created an artificial lake of some size.

The earliest dam still in use today is made of rock and is about 20 feet high, as high as a barn. It is on the Orontes River in Syria and was built more than 3,000 years ago.

After that time, many dams were built by the Assyrians, the Babylonians, and the Persians. Dams also appeared in Ceylon, China, India, and Iran. They were all made of earth or stones. The hard labor of construction was done by slaves.

The ancient Romans built dams and so did the Japanese. During the fifteenth to the eighteenth centuries—the time when our country was still being explored and colonized—dams appeared in Italy, Spain, and France. In 1675 an earthen dam 115 feet high was built at Ferréol, near Toulouse, France. It was the highest earth dam in the world and more than 150 years passed before a higher one was built.

No one worried much about building bigger or better dams for a long time, because there wasn't any special need for them. Dams remained more or less the same for centuries. They were used to form water reservoirs for towns and cities, to deepen

canals for navigation, and to drive water wheels.

The water wheel was an ancient source of power. The Chinese had one of the earliest, the *noria*. The wheel was made of bamboo and was fitted with woven reed paddles or with buckets. It was set in a fast-moving stream and, as the water struck the paddles or buckets, it made the wheel turn. When

the wheel was attached to millstones, it turned the stones and ground grain. The water wheel was at its best when it could be set beneath a waterfall, because the force of the falling water made it turn very quickly.

The ancient Romans were good engineers and they used water wheels widely. They got the idea of building dams to make waterfalls wherever they needed them. Water was stored up high behind a

dam and let out to form a fall when it was time to run the mill.

It must have been considered a wonder when a dam that held enough water for 16 water wheels was built near Arles, France, not long after Biblical times. It turned 32 millstones and produced 30 tons of flour a day. Water mills caught on all over Europe and, by the year 1086, there were more than 5,000 of them in England. Many of them had their own dams.

In England, in the middle of the eighteenth century, power tools started to replace hand tools and factories began to take the place of small workshops. This progress soon spread to France and other parts of Europe. Water power had given industry its start, and now better dams became important so that more factories could be built.

But for a hundred years or so, people didn't seem to learn much that was new about building dams. They tried all of the old ways they had used before; they just built larger dams. Sometimes these worked, but sometimes they fell down.

In the 1850s, W. J. M. Rankine, a professor of civil engineering at Glasgow University, in Scotland, showed that science could alter dam construction. From this time on, dams began to change. Many fine, large dams were built in Europe and in the United States. In 1902, a dam was built across the Nile River, but this time it wasn't a puny dam that gave drinking water to an ancient city. It was a dam that irrigated large areas of useless desert sand and turned it into flowering earth.

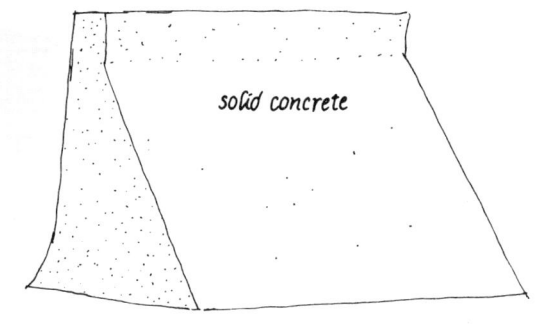

CONCRETE DAM

Three
ALL KINDS OF DAMS

Dams are built of many kinds of materials. Concrete is a particularly good one, but often it isn't practical to cart all of the heavy materials and machinery required to the site. It may be better to build a rock dam if the area is a rocky one. Where lumber is plentiful, a wooden dam is often the best choice. Earth dams, the earliest kind made, are still sometimes the most practical of all. In fact, the

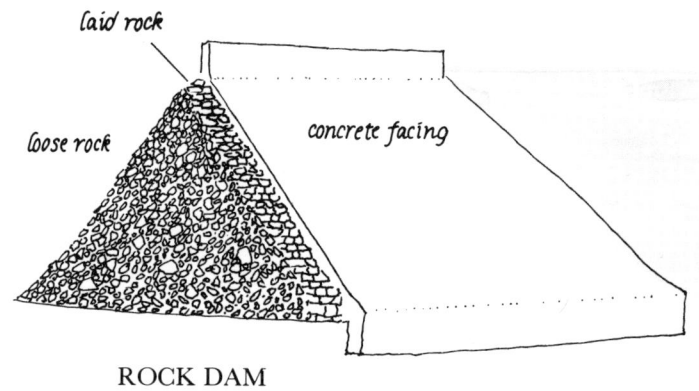

ROCK DAM

WOODEN DAM

largest dams in the world today are all *earth-fill* dams.

These dams are *embankment* dams. To make one, earth, clay, gravel, and the like are banked up in a big heap. The mass is then pressed down with heavy rollers until it is dense and watertight. Finally, it is completely covered over with a close-fitting layer of stone called *riprap*.

Some embankment dams are filled with rocks instead of earth and then they are known as *rock-fill* dams. Often, earth and rocks are used together.

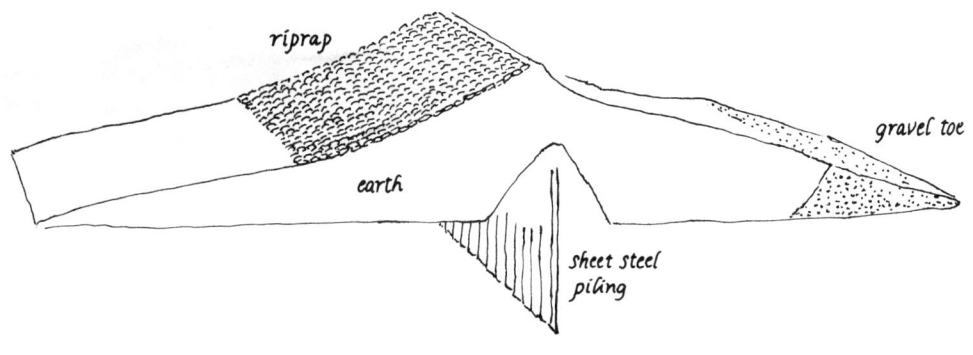

EMBANKMENT DAM

MASONRY DAM

Masonry dams are made of cut blocks of stone or of concrete. These are used to dam narrow streams, like those that run through mountain gorges. They are often extremely high, but not usually very long. For very broad streams, embankment dams are better.

To do the job that best fits the place and the purpose of the dam, concrete dams are made in different ways. Concrete can be pulled apart easily, because, as scientists say, it has little *tensile* strength. But it can support great loads, because it has *compressive* strength. Each dam is designed with this in mind so that it will be as strong as possible.

A *gravity* dam is best to stretch across a fairly wide valley. Its own great weight is what gives it the

GRAVITY DAM

23

strength to hold back the water. A gravity dam usually has a straight wall on the reservoir, or upstream side, and a sloping one on the downstream side. The width at the base is about three quarters the height. The width may be somewhat less if steel rods are anchored to the foundation and embedded in the concrete.

A *buttress* dam is propped up by supporting walls. Since the buttresses take some of the load, the dam doesn't have to be as thick as a gravity dam and so it is cheaper to build.

BUTTRESS DAM

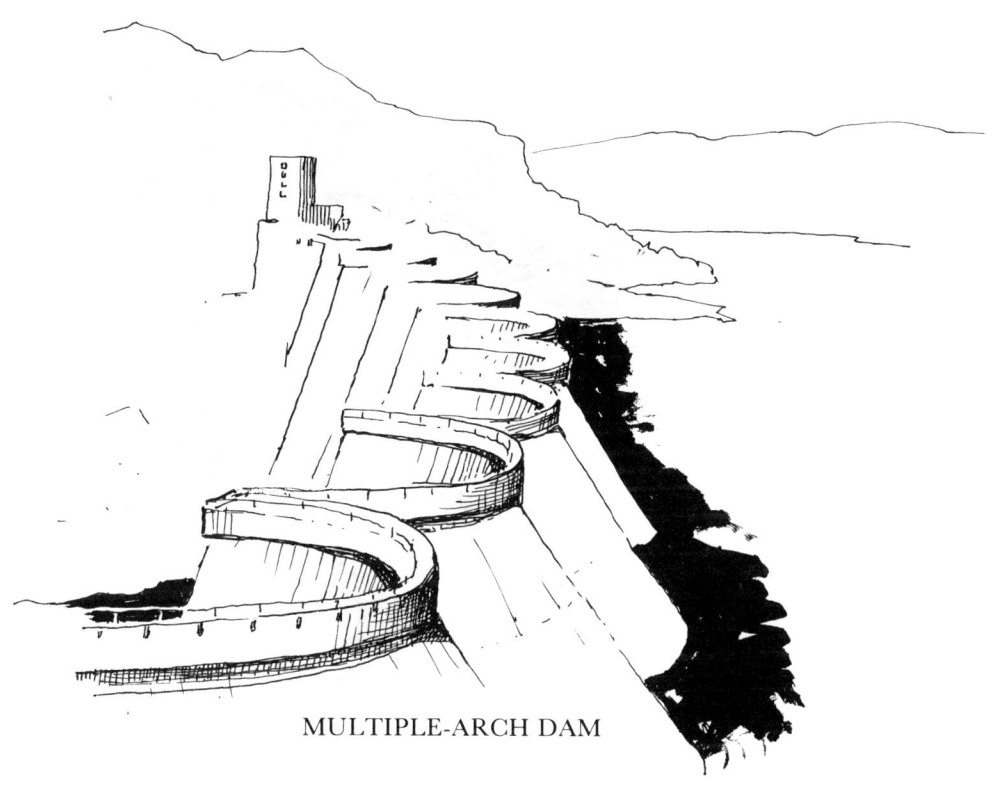
MULTIPLE-ARCH DAM

When a high, narrow dam is needed, an *arch* dam is usually chosen. It is curved outward on the upstream side. The sides of the valley help support the force of the water. Sometimes the dam has more than one arch and then it is called a *multiple-arch* dam.

Arch dams are particularly strong. One such dam was built at Vaiont, Italy, in 1961. It stretched across a mountain gorge and was the second highest dam in the world at 858 feet. On October 9, 1963, there was a huge landslide down the mountain slope. Countless tons of soil and rock fell into the water behind the dam—the reservoir—causing the water to flood over the top of the dam. A rushing torrent poured down the valley to destroy entire villages and drown many people. The dam, most surprisingly, was almost completely unharmed, and is still in use.

To avoid *overtopping*—the flooding of water over a dam—modern dams are provided with *spillways* and *gates*. These allow water to be let out in a controlled and orderly way. In case of unexpected flooding, the newest models open automatically and draw off the extra water in such a way that it cannot do damage. Ancient dams had no spillways and they were almost always destroyed when too much water poured over them at one time.

Because of new engineering features, modern dams can safely be very large.

Four
ELECTRICITY FROM WATER

When factories were first built, water wheels were still used for power. But within a few years, the steam engine—an old invention that had never been very useful—was perfected by James Watt. Now, more factories sprang up everywhere and all of the new machines were made to run on steam power.

For a long time, there had to be a steam engine in every factory. But a big change came when Thomas A. Edison invented the steam electric plant in 1882. Now a steam engine in a single power plant could send electricity through wires to a number of factories, and machines could run on the new power. This was certainly an improvement, but it was only useful to factories close to the power plant. The new electrical power could not travel over long distances.

Then, along came inventor George Westinghouse, who solved that problem. Westinghouse's idea was to use *alternating current,* current that flows back and forth instead of in a single direction. This kind of electrical current could be sent out at very high *voltage,* or force, so that it would travel through wires for many miles. When it arrived at its destination, it could be changed back to low voltage, ready for use.

Even people who lived far from cities could get electricity now, and factories could be built in far-away places. It didn't make sense to build steam electric plants every place, though, because there was often a natural source of power instead: water. Just as falling water had once been used to turn a water wheel, it could now be used to make electricity.

Hydro comes from the Greek word for water, and the first hydroelectric station opened in Niagara Falls, New York, in November, 1896. It sent current to the city of Buffalo, twenty miles away. Today, there is a vast network of hydroelectric plants all over the United States. They produce many millions of kilowatts of energy every year.

In order for water to make electrical energy, it has to fall from a high place to a low one. Sometimes this happens naturally, as at a waterfall or down a mountain slope. But such sources are not always dependable. In dry weather they may slow to a trickle or dry up altogether. As we know, a dam

can store water for dry seasons. It can also make it possible for a hydroelectric plant to be built where there is no natural fall at all. The dam holds back enough water to reach a high level, a *head*. Water cannot make electrical power without a head. When the water is let go, it falls with great force and strikes the blades of giant water wheels called *turbines*.

A turbine is like an old-fashioned water wheel except that instead of turning a mill, it turns a generator. The generator makes electricity. The

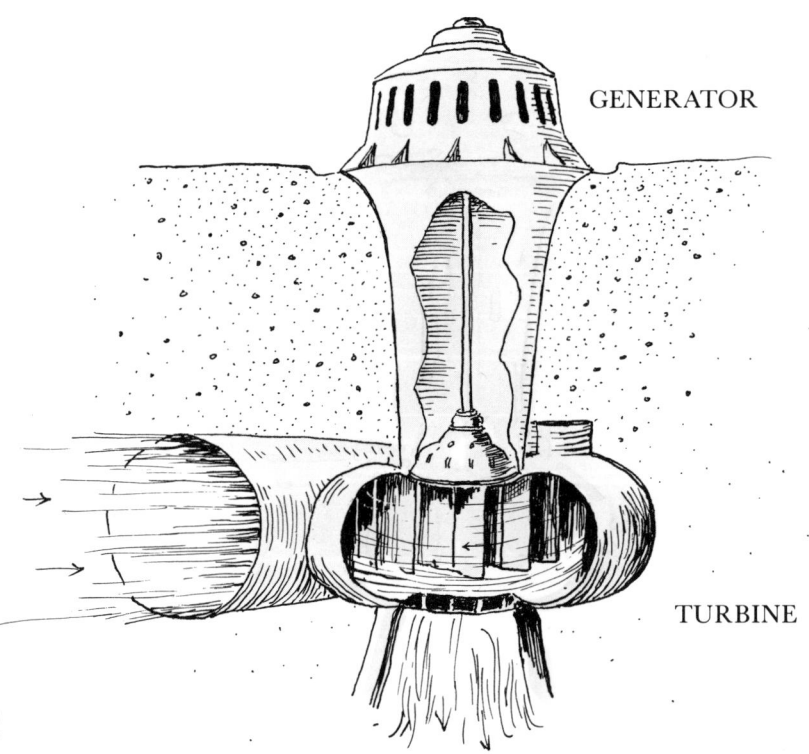

force of gravity from the falling water is the original source of energy which turns the turbine. The mechanical energy of the spinning turbine turns the generator. The generator finally turns the mechanical energy into electrical energy.

The main parts of a hydroelectric project are a dam with its lake, or reservoir, behind it; a pipe, a tunnel, a canal, or something of the sort to carry the falling water to the turbine; a power house—a building to enclose the turbines, generators, and other equipment; and finally, a *tailrace,* a pipe or canal to return the water to the stream.

The greater the distance water falls, the more electrical energy can be made, so hydroelectric plants are usually built in mountainous regions; the natural peaks and valleys make it easy to create a high head. Electric power is made from water power in almost all countries where there are mountain ranges. Hydroelectric stations make nearly one third of all the electric power in the world.

RESERVOIR

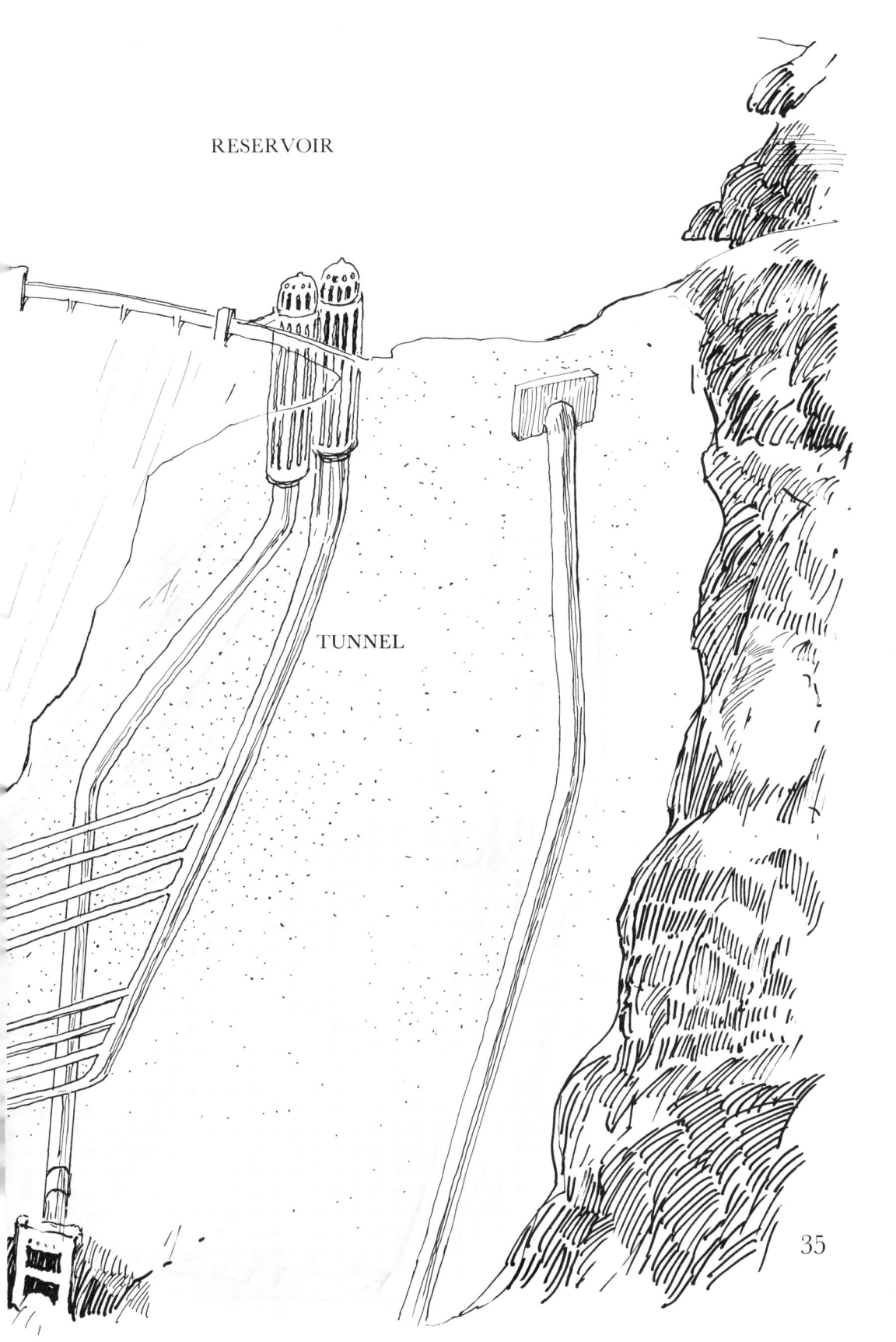

TUNNEL

One of the best-known power systems in the United States was built by the Tennessee Valley Authority; it stretches through much of the South. Among its famous dams are Wheeler and Wilson at Muscle Shoals, in northern Alabama. Some other well-known dams that are part of hydroelectric projects in this country are Grand Coulee, in Washington; Hoover, also known as Boulder, in Arizona and Nevada; Bonneville, in Oregon; Shasta, in California; and Fort Peck, in Montana.

Almost all hydroelectric dams are multi-purpose. In addition to producing power they often supply water to irrigate farmland. They control floods, which, of course, saves homes and lives, but is also important to soil conservation. Whenever farmlands are flooded, there is a danger that the swiftly rushing flood waters will carry off valuable topsoil and ruin the land. Now, the fine farms in many fertile valleys are protected by hydroelectric dams.

A less important use of the dams—but a very

pleasant one—is recreation. In many places, lakes created by dams are enjoyed by millions of people every year. The lakes are beautiful and they are used for swimming, fishing, and boating.

Lake Mead, at Hoover Dam, is a National Recreation Area. It is visited by nearly five million people a year. There is excellent fishing at all seasons and many kinds of boats for hire. The largest ones are cabin cruisers, complete with pilots, and they can be chartered for overnight trips on the lake. The lower part of Lake Mead is used for water-skiing and attracts many visitors from nearby Las Vegas.

Five
PLANNING A DAM

Long before a dam is built, engineers begin to make careful plans. The first thing they do is choose a site for the dam. They have to be sure that it is in exactly the right place, is the right shape, and that the river bottom has the right kind of foundation; it must be either rock, or earth so firm that the weight of a dam will not squeeze it out of shape.

They must figure out where the reservoir will be and how much water it has to hold. They must see what is on the land to be flooded. Often, the reservoir site contains houses, roads, railroads, even entire towns. All of these must be moved, or bought so that they can be destroyed. This is never easy, but once in a while the problems are so unusual that they attract world-wide attention. This was the case with the Aswan High Dam.

The dam was built in the United Arab Republic (Egypt) between 1960 and 1970 and has brought about major changes—mostly good ones—in the economy of the country. It protects the Nile River valley from floods, and irrigates about a million acres of former desert land which can now be used for farming. It also irrigates another 700,000 acres in upper Egypt which used to produce a single crop each year. With its new water supply, the same land now yields up to three crops a year.

Farmers have not been the only ones to benefit from the dam. Its hydroelectric station has given Egypt six times as much electric power as it had before.

But Lake Nasser, the dam's reservoir, is one of the largest artificial lakes in the world. The area it flooded happened to be dotted with priceless temples, monuments, and other ancient treasures.

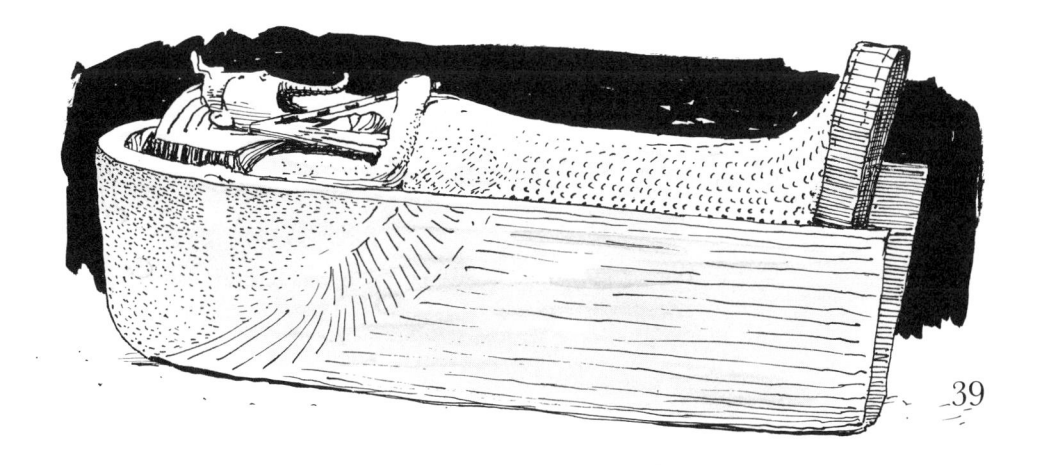

Naturally, it was out of the question to destroy these works, but the job of moving them had to be done by experts and needed painstaking care. Because the United States offered help with the dam project, the Egyptian government made us a gift of the two-thousand-year-old temple of Dendur. It was carefully taken apart and shipped to the Metropolitan Museum of Art in New York City. There it was put together again and is now a major attraction at the museum.

    Moving antiquities is not an everyday problem of dam-building, but moving a river is. A dam can't be built under water and so the river has to be made to flow some place else until the dam is finished. Sometimes large tunnels are built to hold the river. Frequently, it is possible to work on one part of the river bed at a time, while all the water is kept out of the way. This is done by means of

TEMPLE OF DENDUR

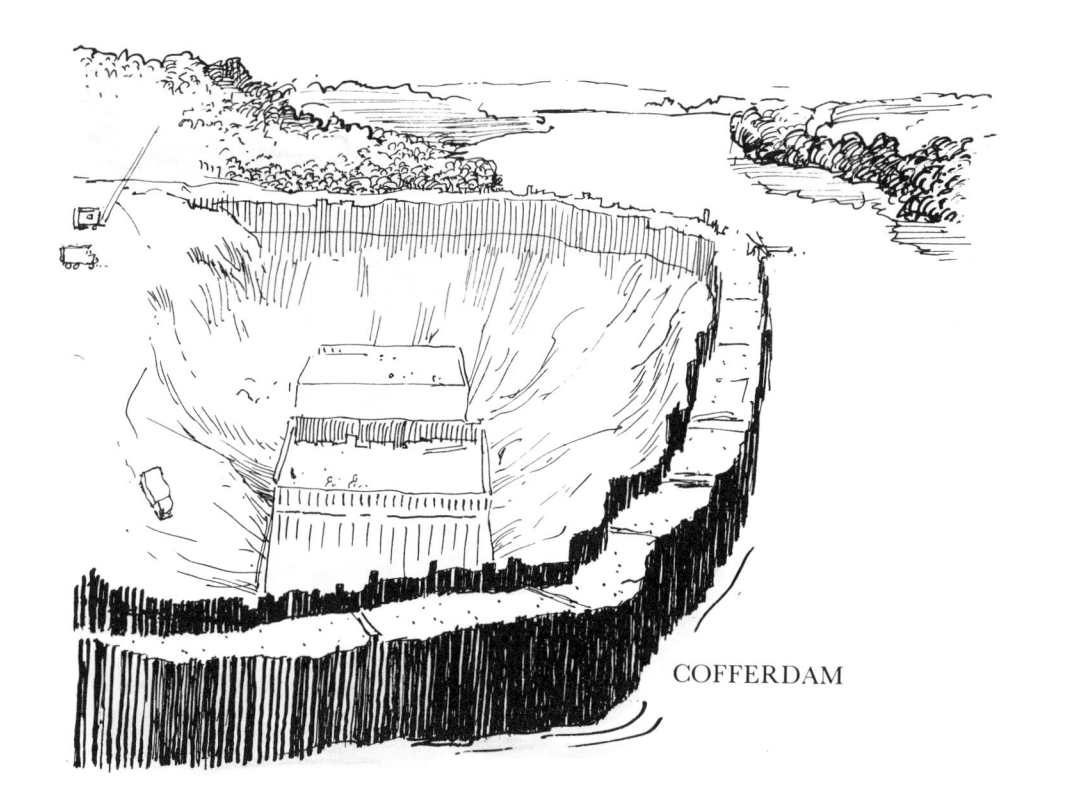

COFFERDAM

cofferdams, small dams that can be moved from place to place as needed. The dams can be set up to form an enclosure which is then pumped dry.

When the question of how to divert the river has been settled, the most important problem remains—which kind of dam to build. Should it be earth, rock, or concrete? Embankment or arch? Then, if the dam is to be used for power, how will the water get to the power house? If the dam is meant for irrigation, or for a city water supply, is there plenty of room for an aqueduct to take the water to where it will be used? A route through mountains or solid rock would not be practical. *Is there a practical route?* How much mud and silt are likely to flow into the reservoir? Too much will shorten the life of the dam.

And then there are problems with migrating fish. In the early days of hydroelectric dams, millions of fish were caught in the turbines. Now, designers save the fish by means of fish passes, a series of pools or locks that allow the fish to pass safely.

Engineers must learn as much as possible about the river and the way it flows, rises, falls, and floods, to be sure that the dam will be strong enough to hold it under any conditions. A dam must be high enough to store all the water needed to do its job. And high enough to store extra water in flood times. But it has to be even higher than that, because at floodtide there might be a wind storm that made big waves. There might be earthquakes or landslides. Every possible event that could cause the dam to be overtopped has to be thought of and allowed for. Accidental overtopping almost always means tragedy, and every effort is made to avoid it.

A well-designed dam has spillways that add to its safety, and they have to be carefully planned, too. *Overflow* spillways are part of the dam itself. *Channel* spillways are built to one side. They may be out in the open in a canal, or enclosed in a chute or tunnel.

Both types of spillways may use gates to help

OVERFLOW SPILLWAY

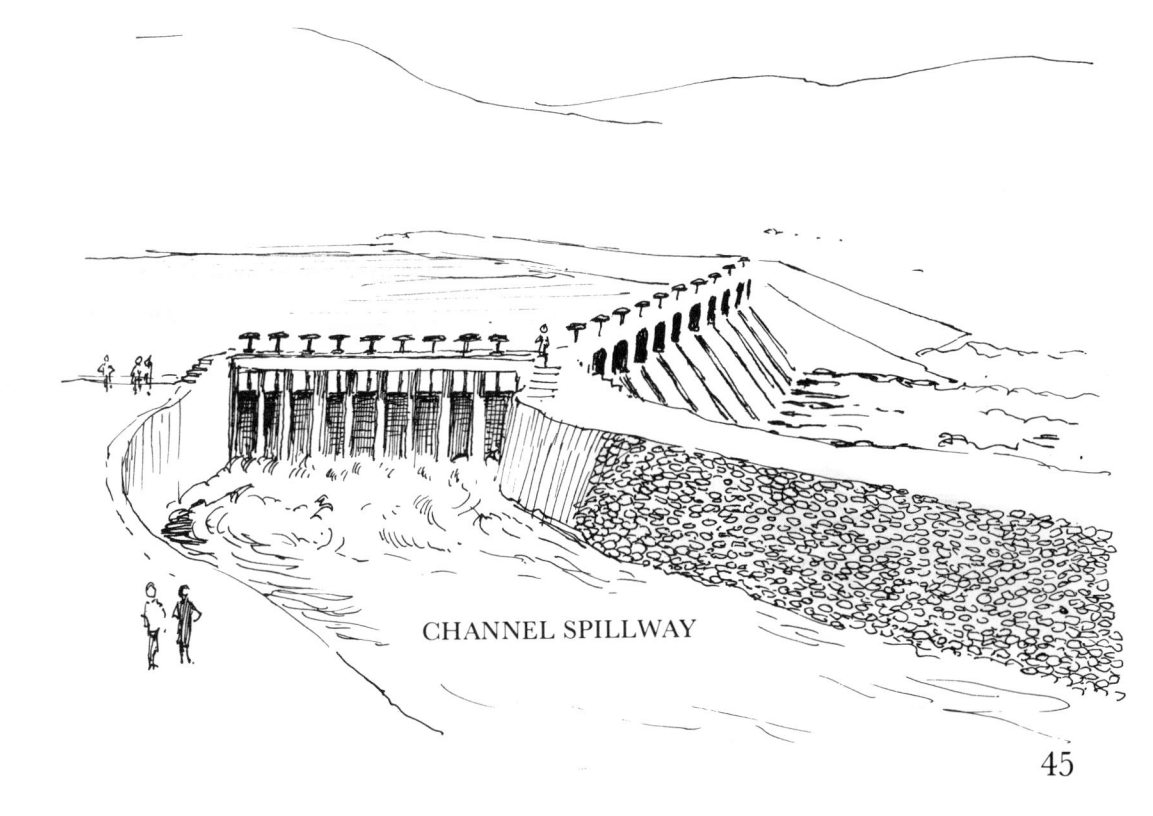

CHANNEL SPILLWAY

VERTICAL-LIFT GATE    RADIAL GATE

control the flow of water. Some of the main kinds of gates are *vertical-lift, radial* or *tainter, tilting,* and *drum.*

Only after every detail has been planned, designed, checked and rechecked, can the real work of building a dam begin. Here is a story about the way one dam—a rather small one by today's standards—was built.

TILTING GATE    DRUM GATE

46

Six

A DAM IS BUILT: THE STORY OF CROTON DAM

During the 1870s, New York City began to need a new water supply. The water it was getting from Croton reservoir in nearby Westchester County would soon be too little. After much discussion, plans for a larger Croton Dam and reservoir were finally started in 1883. Construction began in 1892, and the dam was finished in 1905. To find out why it took thirteen years, let's see what building a dam was like in those days.

First, the engineers studied the course of the Croton River very carefully and picked a site for the new dam. But soon they found out that their choice was no good and so they had to start all over again. They finally decided on a site one mile north of the first one.

Then they began to plan the aqueduct. This would be an enormous tunnel to carry water to the city, more than thirty miles away. The entire tunnel would be lined with stone so that it wouldn't wear out as millions of gallons of water poured through it for years and years. Most of it would have to be deep underground at an average depth of 170 feet.

As soon as the aqueduct was started, the Croton River had to be moved out of the way so that work on the dam could begin. First, the workmen dug a huge channel for the river. Then they put up a masonry wall and two earthen dams to keep the water inside.

At last it was time to start on the dam—and now the problems began to grow greater every day. The foundation was to be built on solid bedrock after a trench had been dug to reach it. As the trench grew deeper, tons and tons of earth were removed. There was so much of it to be hauled away that a system of cable cars had to be installed to do the job.

When the foundation was finally begun, it was in a pit 136 feet deep. Underground springs and leakage from the river flooded it so badly that work had to stop until the pit could be made dry. It was done by a series of steam-driven pumps that pumped out five million gallons of water a day.

The dam design called for a structure 2,200 feet long, 297 feet high, and 200 feet thick at the base. It would be built of huge blocks of granite from a quarry two miles away. The stone—a variety called gabro—was to be hand-cut by stonemasons and carefully fitted into place with cement. The only problem was that there weren't enough skilled stonemasons in the whole United States to build a dam. About 2,000 artisans were brought over from Italy to do the work. They settled in nearby villages and their descendants form a large part of the population of the area today.

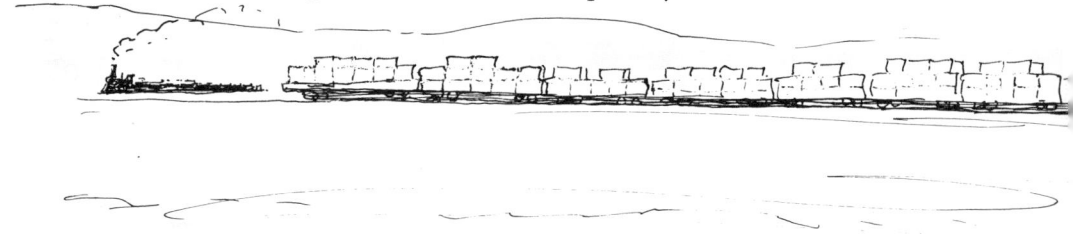

Next, of course, came the problem of how to move the big stones from the quarry to the dam site. In the days before trucks, this was not a simple matter. A special narrow-gauge railway had to be built. Every working day, seven steam locomotives with 83 flat cars puffed back and forth between the quarry and the riverbed.

Each block of stone was carefully washed by the workmen before it was fitted into place with mortar and smaller stones. In the winter, the weather grew so cold that salt was added to the cement to keep it from freezing and it was mixed with water that had been steam-heated. At night, the new work was covered with salt and canvas, and every morning it all had to be washed down. The work went so slowly that the men often labored during the night as well as the day in an effort to get it done.

The reservoir behind the dam had its problems, too. It was going to hold 32 billion gallons of water and would cover an area of land nearly twenty miles long. On that land there were six towns, a number of roads, railroad tracks, farms, churches, and a cemetery. Everything had to be bought or moved. The entire village of Katonah, near Croton, was moved one mile to the south. Roads were rebuilt to zigzag for 75 miles around the reservoir.

The biggest problem of all was the cemetery. No one knew what to do about it. Finally, the dam builders offered a cash reward to all those who would agree to have their ancestors dug up and buried some place else. The plan worked well enough and the Society of Friends moved the bodies that hadn't been claimed.

When the dam was nearly finished, some small cracks were discovered in it. These were repaired, but some last-minute changes had to be made in the design.

At one point, the workers went on strike for a shorter work day and higher wages. They won their strike, but before they did, several regiments of soldiers were called in to keep order.

As the reservoir began to fill, rumors spread through the community that the dam would give way. The rumors were false, but they caused a near panic.

These were only a few of the matters that had to be dealt with. It does not really seem surprising that the dam took thirteen years to build; the surprise is that it didn't take much longer.

When the dam was officially opened, newspapers of the day hailed it as "an engineering marvel," "safe as Gibraltar," and "the second largest hand-made masonry structure in the world, second only to the Great Pyramids of Egypt."

Today, the Croton Dam still sends water to New York City, but now it is only a part of the city's much larger water-supply system. The dam, though, is just as beautiful as it was when it was built. People still come from far away to admire the hand-made masonry and watch as sparkling water cascades over 1,000 feet of spillway to splash into the Croton River 300 feet below.

Seven
WHEN DAMS FAIL

In olden times, it was common for dams to fail. They were not designed scientifically and often turned out to be too weak to bear the weight of water against them. Sometimes they were built on muddy foundations and were easily washed out and overturned. They had no spillways or gates so they were usually overtopped in times of very heavy rain, melting snow, or earthquakes. Earthen dams almost always fell apart when overtopped.

In modern times, dams do not fail *often*, but they do still fail. We have already learned about the landslide at the Vaiont Dam in Italy. Another modern European dam, the Malpasset in southern France, fared far worse and fell apart completely in 1959. A concrete arch dam across a narrow valley, one day it just gave way and poured its whole reservoir of water down into the village of Fréjus. Three hundred people were killed. It was later discovered that the bedrock under the dam was not solid, as had been thought; it had a soft clay seam down one side.

In 1968, a single earthquake damaged 93 embankment dams on the Japanese island of Honshu. Fortunately, none of them was very high, but there was a great deal of destruction anyway.

Following these accidents, engineers have tried even harder to make dams foolproof. The use of models and computer calculations have made dams safer than before, and yet accidents continue to happen. They happen even with the best planning. Sometimes they happen because no one bothered to plan at all.

That was the case at Buffalo Creek, West Virginia. A coal-mining company was in the habit of dumping its "black water"—water that had been used to wash coal—into nearby streams. When ordered to stop polluting the water, the company got another idea. They decided to use waste from the mines, called *slag*, or *gob*, to build dams. In time, the black water would seep through into the stream, but by then, having filtered through the slag, it would be fairly clean.

Buffalo Creek lay in a hollow, a 17-mile-long valley in the Appalachian Mountains. People in the town watched as the slag-heap dam began to grow at the top of the hollow. Finally, there were 80,000 tons of waste on the dam. It was 550 feet wide and 480 feet thick. It had never been properly planned and it had no emergency spillway. Its size, however, was so enormous that it looked as if nothing could ever move it.

One Thursday in February, 1972, a steady downpour of rain began. It rained all day and all the next day. By Friday night, some of the people who lived in the valley were beginning to worry; the water in the reservoir was within a foot of the top of the dam. A few were worried enough to leave, but most just shrugged their shoulders and decided that there was no real danger.

On Saturday morning, February 26, at one minute past eight o'clock, the dam suddenly began to look like melting ice cream as it turned into soft slush. While horrified people stared, it slid forward and 132 million gallons of water poured over its top.

The entire mass of slag and water thundered down the valley destroying everything in its path. It swept away cars, trees, even houses with terror-stricken people hanging on to porch railings or anything else they could grab. The lucky ones saw it coming and had time to scramble up a hillside to safety.

It took three hours for the swirling, zigzagging mass to reach the end of the valley. In that time it killed 125 people and left 4,000 homeless.

This flood, the worst in the history of West Virginia, was the result of pure carelessness. The tragedy can't be undone, but it *can* be hoped that it will serve as a lesson to others.

Dams have served people in many important ways ever since ancient times. And they will continue to make life better for people all over the world, as long as they are planned and built with thought and care.

INDEX

accidents, 26, 58;
  see also earthquakes, landslides, *and* floods
Alabama, dam in, 36
animals, and water, 10
aqueduct, 43;
  for Croton Dam, 48
arch dams, 25, 56
Arizona, dam in, 36
Arles, dam near, 17
Assyrians, and dams, 14
Aswan High Dam, 38-40

Babylonians, and dams, 14
beavers, 5-9, 11
boats, and dams, 10, 37
Bonneville Dam, 36
Boulder Dam, 36
branches, for dams, 5, 11
Buffalo, power to, 30
Buffalo Creek, dam at, 58-62
buttress dams, 24

Cairo, dam near, 13
California, dam in, 36
canals, dams and, 15
Ceylon, dams in, 14
China, dams in, 14
Chinese, and norias, 15
cofferdams, 42
concrete, for dams, 11, 20, 22, 23, 56
conservation, of soil, 36
Croton Dam, building of, 47-55
Croton River, 47, 48, 55
current, alternating, 30

dams, first, 5, 8, 11;
  uses of, 6, 9-11, 14-19, 32, 33, 36-37;
  for recreation, 10, 36-37;
  multi-purpose, 11, 36;
  ancient, 13-19;
  increased need for, 18-19;
  improvements in, 19, 58;
  kinds of, 20-26, 42;
  for hydroelectric plants, 32, 33, 36, 39;
  planning for, 38-46;
  building of, 47-55;
  failing of, 56-62;
  models of, 58
Dendur, temple of, 40
desert, dams and, 19

drum gates, 46

earth, for dams, 5, 11, 14, 20-21, 56
earth-fill dams, 21
earthquakes, 44, 56, 58
Edison, Thomas A., 29
Egypt, dams in, 13, 39-40
electricity, 29-33, 39;
  steam for, 29-30;
  water for, 30-36
embankment dams, 21, 22, 58
energy, *see* power
England, water mills in, 17;
  power tools in, 18
Europe, water mills in, 17;
  power tools and factories in, 18;
  improved dams in, 19

factories, 18, 28-30
Ferréol, dam at, 14
fish, migrating, 43
floods, prevention of, 10, 36;
  dams and, 13, 26, 39, 44, 57, 61-62;
  see also land, flooding of
Fort Peck Dam, 36
France, dams in, 14, 17, 56-57;
  power tools and factories in, 18
Fréjus, flooding of, 57

gates, 26, 56;
  kinds of, 46
generators, 32-33
grain, grinding of, 16, 17
Grand Coulee Dam, 36
gravity dams, 23-24

head, 32, 33
Honshu, dam at, 58
Hoover Dam, 36
hydroelectric plants, 30-31, 39;
  sites for, 31-33;
  parts of, 33;
  and fish, 43

India, dams in, 14
Iran, dams in, 14
irrigation, dams for, 19, 36, 39
Italy, dams in, 14, 25-26, 56;
  stonemasons from, 50

Japan, and dams, 14, 58

Katonah, moving of, 53

63

Lake Mead, 37
Lake Nasser, 39
lakes, dam-created, 10, 13, 37, 39
land, flooding of, for dams, 8, 11, 38, 39
landslides, 26, 44, 56
lodge, of beavers, 6

masonry dams, 22, 54-55
Malpasset Dam, 56-57
Memphis, Egypt, 13
Menes, King, 13
Metropolitan Museum of Art, temple at, 40
millstones, and water wheel, 16, 17
Montana, beaver dam in, 6-7; dam in, 36
mud, *see* earth
multiple-arch dams, 25
Muscle Shoals, dam at, 36

Nevada, dam in, 36
New York City, water for, 47, 54
Niagara Falls, 30
Nile River, dam on, 13, 19; and Aswan Dam, 39
noria, *see* water wheels

Oregon, dam in, 36
Orontes River, dam on, 14
overtopping, 26, 44, 56, 61

people, and water, 10
Persians, and dams, 14
plants, and water, 10
power, 10;
 water wheel for, 15;
 water, 18, 28, 30-36;
 steam, 28-30;
 electric, 29-36, 39
power tools, 18

radial gates, 46
Rankine, W. J. M., 19
recreation, dams and, 10, 37
reservoirs, 14;
 placing of, 38;
 for Croton Dam, 47, 53
riprap, 21
rivers, moving of, 40, 42, 48
rock-fill dams, 21
rocks, *see* stones

Romans, and dams, 14; and water wheels, 16

Sadd-el-Kafara, dam at, 12
Schweinfurth, Georg, 13
Shasta Dam, 36
slag, for dams, 58-59
slaves, and dams, 14
Spain, dams in, 14
spillways, 26, 56, 59;
 overflow, 44;
 channel, 44
steam engine, 28-29
stones, for dams, 5, 11, 13, 14, 20-22; for Croton Dam, 48, 50-51
Syria, dam in, 14

tainter gates, 46
Tennessee Valley Authority, 36
tilting gates, 46
Toulouse, dam near, 14
turbines, 32-33

United Arab Republic, *see* Egypt
United States, improved dams in, 19; hydroelectric plants in, 31, 36; and Aswan Dam, 40

Vaiont, dam at, 25-26, 56
vertical-lift gates, 46
voltage, 30

Washington, dam in, 36
water, holding back of, 6, 9, 10, 11, 32;
 storing of, 9, 32, 44;
 need for, 10;
 for power, 18, 28, 30-36;
 for irrigation, 19, 36, 39;
 quantity of, 38;
 for New York City, 47, 54;
 black, 58;
 see also floods
water mills, 16-17
water wheels, 15-17, 28, 32
waterfalls, dams for, 16
waterways, 10
Watt, James, 28
West Virginia, dam in, 58-62
Westinghouse, George, 30
Wheeler Dam, 36
Wilson Dam, 36
wood, for dams, 20